NEXT-GENERATION MILITARY TECHNOLOGY

ADVANCED WEAPONRY

IN THE MILITARY

by Cynthia Kennedy Henzel

BrightPoint Press

San Diego, CA

BrightPoint Press

an imprint of ReferencePoint Press, Inc.
Printed in the United States

For more information, contact:
BrightPoint Press
PO Box 27779
San Diego, CA 92198
www.BrightPointPress.com

LIBRARY OF CONGRESS CATALOGING-IN-PUBLICATION DATA

Name: Henzel, Cynthia Kennedy, author.
Title: Advanced weaponry in the military / by Cynthia Kennedy Henzel.
Description: San Diego, CA: ReferencePoint Press, 2027 | Series: Next-generation military technology | Includes bibliographical references and index. | Audience: Grades 7–9
Identifiers: ISBN 9781678214180 (hardcover) | ISBN 9781678214197 (eBook)
The complete Library of Congress record is available at www.loc.gov.

CONTENTS

AT A GLANCE

- New nonlethal weapons protect soldiers and civilians. The military is using new technology to develop weapons faster.
- Artificial intelligence is used to automate weapons and make combat decisions.
- The US Army is replacing the Bradley Fighting Vehicle with the XM-30. The new technology can make the XM-30 an uncrewed vehicle.
- The Precision Strike Missile will provide the United States with the ability to hit targets at a much greater distance than older missiles.
- The B-21 stealth aircraft is replacing the B-2. The B-21 is smaller than the B-2 and can carry less weight, but it costs much less than the older aircraft.

- Drones, and the ability to defend against drones, is vital new technology.

- The Aegis Combat System, used on US ships, is one of the most advanced weapons systems in the world.

- The United States is developing a fleet of uncrewed ships and submarines to support the US fleet.

LASER GUNNER

Kelsey Atherton was at a firing range in New Mexico. He was writing a story on new technology. Atherton had a laptop. He held a **controller**. It looked like the controller on his old Xbox. But this system was not for games.

A small vehicle was parked on the firing range. Mounted on its back was a ball-shaped device. It produced a laser. This laser could melt through a drone.

Remote-controlled High Energy Laser Weapon Systems provide one way to counter the drones used in modern military operations.

A laser weapon beams intense energy to a target over long distances. The laser uses the heat of electromagnetic energy to burn or damage targets.

Atherton was testing how well the weapon worked.

The laptop screen displayed four sections. They showed different views of the area. They also showed a map. A red line marked the safe zone. This was a protected area. The trigger would not work if he tried to fire into this area.

Atherton flipped a switch to turn on a laser pointer. This identified the target. A second switch armed the laser. The destructive laser did not work well for targeting. It was so bright that the gunner could not see the target.

A drone appeared on the screen. Atherton moved the controller's left joystick. This placed a tracker marker next to the drone. The ball-shaped laser device would now rotate to follow the drone. Atherton moved the right joystick. This placed the crosshairs on the drone. Then he held down the trigger. The laser fired. The drone's rotor quickly melted. It crashed to the ground.

An aboveground tunnel at the Naval Surface Warfare Center in Dahlgren, Virginia, was used to test weapons during World War II (1939–1945). It has been transformed to test high energy laser weapon systems.

DEFENDING THE COUNTRY

Atherton was using a High Energy Laser Weapon System (HELWS). Lasers are one of many advanced weapons that the

US military uses for defense. On land, new vehicles move troops and equipment faster than ever. In the air, new airplanes and drones protect the skies. Other advanced weaponry keeps the seas safe.

There are also new **nonlethal** weapons. These protect US soldiers. They also prevent civilian deaths. The United States spends more on its military than any other country.

CHAPTER ONE

STAYING AHEAD

In 2024, the United States spent $997 billion on its military. China was the second-highest spender. It spent $314 billion. Militaries must pay troops and buy weapons. It takes money to operate bases. Militaries research new technology, too. In 2024, the US military spent $138 billion on this work.

The job of defending the United States is always changing. New weapons require

In the military, drones, or unmanned aerial vehicles (UAVs), are used for everything from gathering information to precise strikes on targets.

new defense methods. These new weapons include drones. Annabel Flores works for Raytheon. This company makes military technology. Flores says, "Drone threats are particularly dangerous in any environment—from soccer stadiums to battlefields—because they're not easy to spot and take down."[1]

Military Spending

The United States spends more money on the military than other countries do. But the United States is a large country. Some smaller countries actually spend more per person. This includes the United Arab Emirates. It also includes Qatar, Oman, and Saudi Arabia. North Korea spends 23.5 percent of the country's total budget on the military. That is more than twice the budget of the second-highest spending country.

The military is concerned about drones. But other new technologies pose threats, too. In 2025, Russia claimed to have a new nuclear cruise missile. Cruise missiles fly long distances to hit targets. Russia also said it had an uncrewed nuclear submarine. If true, these could be serious threats. North Korea has advanced weapons, too. It has missiles that can reach the United States. China has new ships and airplanes. It is also creating new cyber weapons. These could destroy communication systems or utilities. The US military must defend itself against these threats. This is a big job.

A NEW APPROACH

In the past, developing a new weapon could take years. But technology is changing fast.

This means the military must move faster. In April 2025, Defense Secretary Pete Hegseth made an announcement. He said the military was adapting to meet new challenges. This plan includes new technologies. The military is working to develop weapons faster.

For example, in 2010, the US Army decided it needed a new rifle. The new gun would replace the old M4. The army wanted a rifle that would shoot farther. It also wanted enough power to shoot through armor. It asked companies to bid on the project. No company offered a rifle the army liked. So, the army cancelled the bid. The army tried again in 2017. Several companies offered new models in 2020. SIG Sauer won the bid in 2022.

In 2019, the US gun manufacturer SIG Sauer had contracts to supply guns for police and military in eighty countries.

The M7 finally arrived in 2025. But by then, the army did not need some of the new features. The M7 can fire accurately up to 660 yards (1,600 m). That is farther than most soldiers need to fire. The M7 can also pierce body armor. But the enemies most soldiers will face do not wear body armor anymore. Plus, the M7 weighs 15.4 pounds (7 kg). The M4 weighed only 8.5 pounds (3.9 kg). The M7 is a good rifle. But the battlefield has changed.

Anti-drone systems include turrets atop military vehicles and handheld devices.

The military has a new way to develop advanced weaponry. It uses **simulation**. In 2025, the Army Research Lab worked with Duality. This company developed a new anti-drone system. The system spots drones. Then it warns soldiers of the threat. The team created a **virtual** model of the system for testing. This meant the army could make changes before building the system. Apurva Shah is a cofounder of Duality. He explained that this new system is changing the way military technology will be created in the future.

NEW TECHNOLOGY

The US military creates the most advanced weaponry in the world. This includes air-to-air missiles. These are missiles that planes fire at other planes. One of the latest is the Joint Advanced Tactical Missile (JATM). Development began in 2018. The US military was testing it in the early 2020s.

Little is known about the secretive JATM program. But experts believe the missile will fly much farther than earlier weapons. **Stealth** fighter jets will carry the missiles.

Autonomous vehicles are another new technology. Autonomous vehicles can go to areas that are dangerous for soldiers. The Defense Advanced Research Projects Agency (DARPA) began developing self-driving vehicles in the early 2000s. It held events called DARPA Grand Challenges. These were contests for self-driving vehicles. Some were for off-road driving. Others were for city driving.

In October 2023, US Marines tested an Expeditionary Modular Autonomous Vehicle at the Marine Corps Air-Ground Combat Center in Twentynine Palms, California.

The military uses AI technology to check the location of military teams in the field.

This research helped create the self-driving cars used today. The advances spread to military vehicles, too.

Another major part of technology is artificial intelligence (AI). AI is the ability of a computer to do things that usually require human intelligence to do. In December 2024, the military started a $100 million AI program. AI will help leaders make decisions faster. It will also improve the safety of autonomous vehicles. AI will make weapons more accurate as well.

CHAPTER TWO

ON THE GROUND

The Bradley M3 is a fighting vehicle. It has been used for combat since 1981. The M3 has been updated many times. The newest version weighs 80,000 pounds (36,000 kg). It has a 1 inch (25 mm) cannon. It also has a large machine gun. And it has anti-tank missiles. Geoff Norman is the director of US strategy at General Dynamics Land Systems. He says of the

The US Army used the Bradley Fighting Vehicle to begin testing features that would be used on the XM-30. The XM-30 can be operated remotely, or it can be manned.

US soldiers who are members of Stryker Brigade Combat Teams train at the National Training Center in Fort Irwin, California.

Bradley, "It's at the point where it can't be upgraded further."[2]

The army is replacing the M3. The new vehicle is called the XM-30 Combat Vehicle. The vehicle can run with a crew. But it can also be controlled remotely. The XM-30 is designed to be easy to upgrade with new technology as it is invented.

The XM-30 has a 2 inch (50 mm) cannon. Its machine guns use an infrared sensor. This makes it easier to find targets under poor conditions. Its firing system is supported by AI. The firing system can automatically find and track targets. Then it suggests the best options to take out the target. The XM-30 has armor that is attached in pieces. This reduces weight. Plus, damaged armor can be easily replaced.

For fast **deployment**, the army has the Stryker. Strykers are eight-wheeled combat vehicles. They are lighter and easier to move than vehicles like the XM-30 that move on tracks. Strykers carry a commander, a driver, and nine troops. They can travel up to 60 mph (100 kmh) on roads. Strykers can be adapted for many different purposes. Some are built for medical **evacuations**. Others have special features for **reconnaissance** or combat.

TAKING DOWN THE ENEMY

In 1987, the United States and the Soviet Union signed the Intermediate-Range Nuclear Forces (INF) Treaty. This agreement was an effort to reduce the number of nuclear weapons. It banned missiles that could travel between 310 and 3,420 miles (500 and 5,500 km). But in 2019, the United States withdrew from the treaty. Russia was building new missiles with a range of 1,600 miles (2,500 km).

The United States responded to the threat. It produced the Precision Strike

PrSM can hit targets greater than 250 miles (400 km) away.

Missile (PrSM). The army received the first 26 missiles in 2024. It planned to receive 400 more by the end of 2027. Newer models will have a range of 620 miles (1,000 km).

The PrSM has other new features. For example, it can hit moving ships. Hitting a ship from far away is hard. It takes time to lock on a target and fire a missile. In that time, the ship might move. The PrSM has

Overkill

In 2017, a Patriot missile was used to shoot down a drone. It worked. But Patriot missiles cost about $3 million each. A drone costs a few hundred dollars. General David Perkins pointed out the need for anti-drone technology. Shooting down drones would soon use up all the Patriots.

new technology that helps track moving targets. The PrSM is also thinner than older missiles. This means two missiles can be loaded into a launcher at the same time. The missiles can then be fired more quickly.

PrSMs work well against distant targets. But they cost $3.5 million per missile. This means they are not practical for new threats such as drones. Lasers help meet this threat.

The HELWS laser can stop a swarm of drones. The laser fires through a lens. The lens focuses the beam to point in a particular direction. Different power levels shoot the weapon to different distances. It has a maximum range of 1.8 miles (3 km).

The laser does not require bullets. It is easy to move. Plus, it is cheap to operate.

This is helpful when defending against inexpensive drones. Annabel Flores of Raytheon says, “It makes no sense to shoot . . . a million-dollar missile into something that’s a thousand dollars.”[3]

NONLETHAL SYSTEMS

Not all defense happens on a battlefield. Sometimes, airplanes or ships move too close to US borders or ships. Terrorists hide in cities. In these cases, shooting a missile is not possible. Nonlethal weapons are an option. They allow the military to control these situations without harming people.

The Active Denial System is used to control crowds. It uses a beam of energy. The energy enters the skin to a depth of about three sheets of paper. The energy

The Active Denial System was available during the Afghanistan War (2001–2021), but it was not used.

In October 2025, marines received training in the use of nonlethal weapons at the Marine Corps Base Camp Pendleton in California.

makes the person feel as if their skin is burning. The person can step out of the energy beam to avoid harm.

Other nonlethal weapons include lasers that temporarily blind enemies. Underwater, the Impulse Swimmer Gun creates a sound wave. It makes swimmers within 160 yards (150 m) feel sick. The RF Vehicle Stopper uses microwaves to shut off vehicle engines. The same technology could be used from airplanes to stop small ships.

IN THE AIR

The first B-2 stealth bombers flew in 1989. They are famous for their flying-wing shape. The B-2 is being replaced with the B-21 Raider. The B-21 wingspan is about 40 feet (13 m) shorter than the B-2's. It can carry 30,000 pounds (13,600 kg). This is half the load of a B-2. But it will be less than half the cost of a B-2.

Stealth is an important feature. On radar, the B-2 appears to be the size of a bird.

Flight tests for the B-21 were conducted at Edwards Air Force Base in California.

AF 0001

But the B-21 looks even smaller. New materials and design make this possible. The B-2 has stealth features only on the front. The B-21 has stealth features around the entire plane. The engine exhaust is less detectable on the B-21. This makes it harder for heat-seeking missiles to find. The technology behind these improvements is still secret.

No Co-Pilot

The B-2 has two seats for a pilot and co-pilot. There are also two seats in the B-21. One is for the pilot. The other seat is for a weapons officer. A co-pilot is not needed. This is because AI is built into the plane. Eventually, the B-21 may be able to fly uncrewed.

The B-21 is a long-range bomber. But it is also built for reconnaissance and electronic warfare. Electronic warfare includes jamming, or blocking an enemy's electronic systems. The B-21 has AI. It helps make decisions. AI analysis of whether to attack, jam, or avoid a threat could save a mission.

DRONES

Drones have become a key part of modern warfare. The US military is developing a fleet of uncrewed combat aerial vehicles (UCAVs). Daniel Driscoll is the US Army Secretary. In 2025, he said, "We expect to purchase at least a million drones within the next 2 to 3 years."[4]

Drones are cheaper and faster to build than fighters flown by people. They can be made for specific missions. For example, Gambit drones begin with a common base. Then parts are added to complete a particular mission. Gambit 1 is used for reconnaissance. It can lead a strike and provide early information. Gambit 2 has air-to-air weapons. It can join a fight with crewed fighters. Gambit 3 is made just for training. It helps soldiers learn how the drones work with US systems.

Gambit 4 is made for long stealth missions. It can spy deep behind enemy lines. Gambit 5 is made to work with ships. Gambit 6 is built for air-to-ground missions. It can work alone. It can also work with a group to take out targets.

A life-size model of the Gambit UCAV was displayed in Dubai in 2025. The five Gambit craft have different specialties, including surveillance, intelligence, and combat.

The United States is also learning new ways to defeat enemy drones. There are several ways to detect drones. Infrared sensors see heat from drones. Radio frequency sensors pick up a drone's

A Drone Buster (front) is often paired with a smart shooter (back). The Drone Buster jams the drone's signal while the smart shooter digitally locks in the target before allowing the gun to fire.

radio signals. New technology even identifies drones by the sound they make.

There are several types of anti-drone weapons. Some systems capture drones with nets. Other weapons use microwaves to disable drones. This includes Mjölnir. The weapon is named after the hammer of the Norse god Thor. Mjölnir detects approaching drones. It tracks them. It then uses a burst of microwave energy to disable them.

Some attacking drones are made from cheap drones that anyone can buy. Silent Archer is used to defend against these small, slow drones. Silent Archer can detect and track them. It can also identify the type of drone. It then jams communications between the controller and the drone.

The Drone Buster is another anti-drone weapon. It weighs about 6 pounds (3 kg). This gun-like weapon can be held by hand. The Drone Buster disrupts the signal between the drone and the controller. It can also disrupt satellite signals sent to automated drones. New models fool drones

The US Army selected Bell Helicopter, which developed the Bell V-280 shown here, to develop the MV-75. The aircraft is expected to be ready for use in 2031.

into flying away. This provides soldiers with a larger safety zone.

STRAIGHT UP

The US military uses helicopters in combat missions. Current craft include the Apache and Black Hawk. The MV-75 is a new helicopter. It is also known as the Future

Long Range Assault Aircraft. "It will provide the speed, range, and endurance needed to conduct air assault, [medical evacuation], and resupply missions," says Brigadier General David Phillips.[5]

The MV-75 helicopter flies faster than the Black Hawk helicopters. It is said to go double the distance at twice the speed. It also has a greater range and can maneuver better.

The MV-75 has a tiltrotor. This is a rotor that can move in different directions. It allows the MV-75 to fly like an airplane. Or it can go straight up like a helicopter. All these new technologies will keep US military air power strong in the future.

AT SEA

The Aegis Combat System is one of the most advanced weapon systems in the world. It forms a protective bubble around a ship. It senses incoming threats. Then it responds to the threat. Aegis combines computers, sensors, weapons, communications, and people into one system. It can track more than 100 targets. These may come from the air. They may

The mast of this destroyer ship holds an Aegis Combat System, which is produced by Lockheed Martin. In 2022, 110 ships were equipped with the system.

Some submarines are powered by nuclear energy. This allows them to stay submerged for months at a time.

also come from the sea's surface or from submarines.

The Aegis system was first deployed in 1983. It has been updated many times. Ships can deploy the proper system for the type of ship. They can even change which weapons or sensors they need for each mission.

Aegis software can also help make decisions. Rich Calabrese works with US Navy systems. He provides an example of a situation a commander might face. "Do I shoot a long-range weapon at this thing and take it out early? . . . Can I wait? If I wait, what [weapon] am I using?"[6]

Aegis can provide information to help the operator choose the best action. This includes choosing which weapon to use for a target. But operators can still override the system if needed.

AIRCRAFT CARRIERS

Aircraft carriers are some of the largest military ships. They can launch and land dozens of aircraft. Navies use them to extend their fighting ability over a wide area. Carriers also help deliver aid after natural disasters.

In September 2005, the aircraft carrier USS *Harry S. Truman* delivered bottled water and other relief supplies to the victims of Hurricane Katrina in New Orleans, Louisiana.

PARTS OF AN AIRCRAFT CARRIER

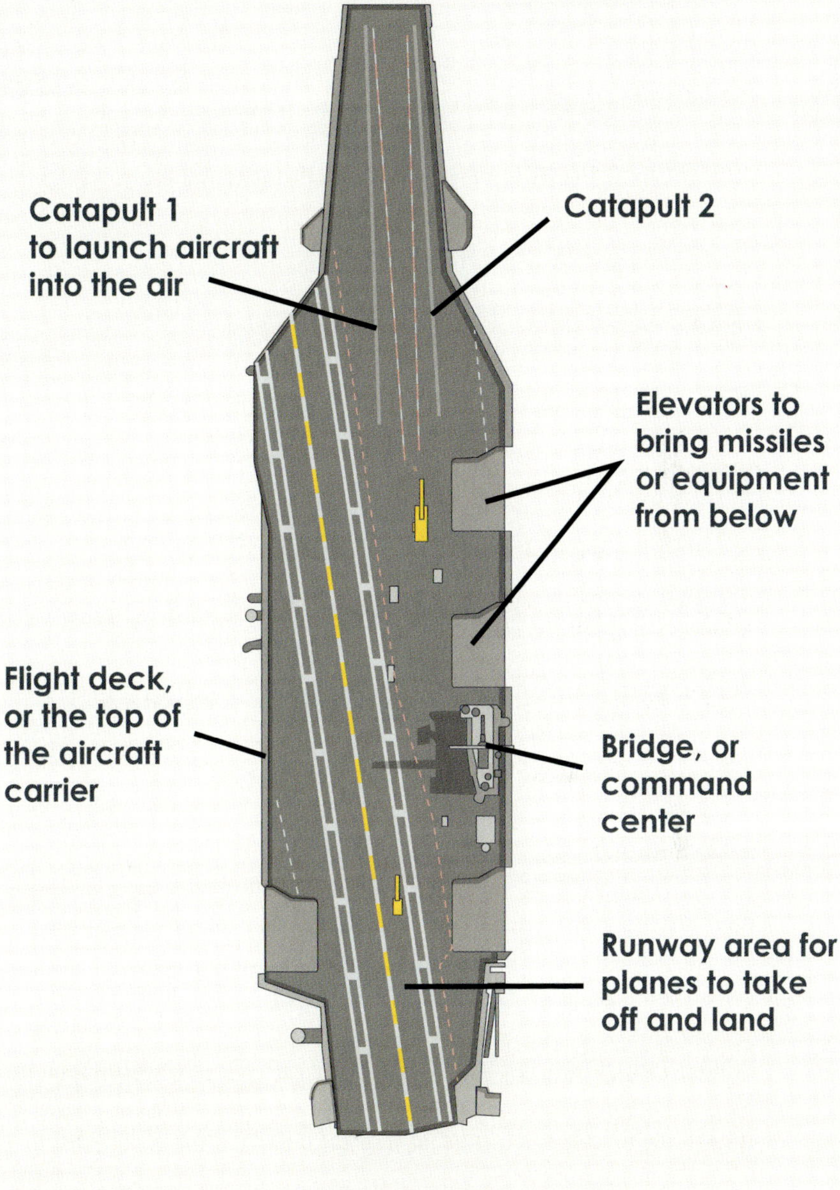

An aircraft carrier is a mobile airport for military aircraft. The catapults thrust aircraft 300 feet (90 m) into the air during takeoff.

The largest carrier in the world is the USS *Gerald R. Ford*. It is nuclear powered. This means it needs refueling only once during its 50-year life. The USS *Gerald R. Ford* was deployed to the Caribbean Sea in May 2025.

The USS *Gerald R. Ford* uses a new technology to launch aircraft. Older ships use steam to power catapults. Catapults are the machines that launch aircraft from the ship. The *Ford* has an electromagnetic launch system. This system uses powerful electromagnets to pull the aircraft forward. The ship does not have to produce steam. And the crew has more control over how much force is used. This means smaller planes can use the launch system without getting damaged.

The USS *Gerald R. Ford* first deployed from Norfolk, Virginia, on May 2, 2023. It is the largest warship ever built.

GHOST FLEET

Uncrewed ships, or USVs, will soon be common. They can carry materials such as ammunition to supply crews. They can also fire weapons. The larger USVs fire missiles at sea and land targets. The smaller ships place explosives to destroy enemy submarines. USVs can be controlled remotely. They can also function on their own. USVs are cheaper, safer, and smaller than crewed ships.

The *Sea Hunter* was the first USV. It was the first warship to use AI in place of a crew. The *Seahawk* was the second USV. It is 135 feet (41 m) long. Dan Brintzinghoffer is a ship systems expert. He says, "*Seahawk*

Uncrewed Subs

DARPA's *Manta Ray* is an uncrewed submarine. It is shaped like the B-21. It is made for long undersea missions, as it is able to go where people can't. It glides at low power, making it hard to detect. *Manta Ray* uses new technologies for power. These include energy produced by the difference in ocean temperature at different depths.

In March 2024, the DARPA *Manta Ray* was tested at the Naval Base Ventura County Point Mugu Sea Range in California.

is unique because it was designed from the keel up to be unmanned for weeks or months at a time. We didn't just put an autonomous navigation system onto an existing ship."[7]

Other USVs are larger. The *Mariner* and *Ranger* are 193 feet (59 m) long. *Mariner* has a small version of the Aegis system. It can work as a mothership for other USVs.

These large USVs carry a small crew that can take control in an emergency.

In 2025, DARPA began at-sea testing of the *Defiant*. It will provide less expensive support to the US Navy's fleet. Greg Avicola works for DARPA. He says, "*Defiant* is a tough little ship and defies the idea that we

The USV *Mariner* (top) cruises the Pacific Ocean alongside USV *Ranger* in September 2023. In January 2026, the navy had plans to create three USV divisions.

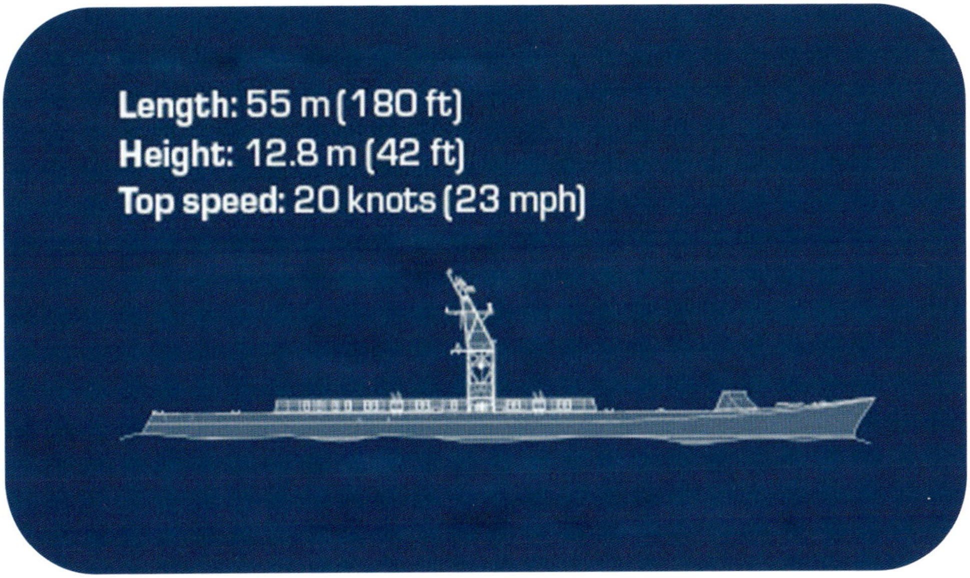

The *Defiant* is specifically designed to operate without crew members. The ship does not even have halls for people to walk through.

cannot make a ship that can operate in the challenging environment of the open ocean without people to operate her."[8]

Technology gives the military the tools to succeed on land, in air, and at sea. It is quickly moving to build uncrewed weapons. But it takes people to improve technology. It takes people to build and maintain the weapons. People are also needed to use the weapons effectively.

GLOSSARY

autonomous vehicles

vehicles with sensors and AI software that are able to operate with little to no human help

controller

a device used to control action on a computer screen

deployment

moving troops or equipment to a place for military action

evacuations

quickly leaving a place, usually for safety reasons

nonlethal

will not cause death or serious injury

reconnaissance

observing an area to locate enemies or strategic features

simulation

using a computer to model or study something

stealth

technology that makes detection by radar difficult

virtual

when something appears to exist through the use of computer software

SOURCE NOTES

CHAPTER ONE: STAYING AHEAD

1. Quoted in Inder Singh Bisht, "Raytheon Integrates Laser Counter Drone System with NASAMS Air Defense," *The Defense Post*, April 6, 2022. https://thedefensepost.com.

CHAPTER TWO: ON THE GROUND

2. Quoted in Colin Demarest, "General Dynamics Eyes Summer Delivery of XM30 Prototype," *Axios*, October 15, 2025. www.axios.com.

3. Quoted in Kelsey D. Atherton, "What It's Like to Fire Raytheon's Powerful Anti-Drone Laser," *Popular Science*, October 31, 2022. www.popsci.com.

CHAPTER THREE: IN THE AIR

4. Quoted in Phil Stewart and Idrees Ali, "Exclusive: US Army to Buy 1 Million Drones, in Major Acquisition Ramp-Up," *Reuters*, November 7, 2025. www.reuters.com.

5. Quoted in PEO Aviation, "Army Designates MV-75 as Mission Design Series for Future Long Range Assault Aircraft," *US Army*, May 27, 2025. www.army.mil.

CHAPTER FOUR: AT SEA

6. Quoted in Tyler Rogoway, "Everything You Ever Wanted to Know About the Navy's Ever-Evolving Aegis Combat System," *TWZ*, March 4, 2021. www.twz.com.

7. Quoted in Brandon Buckner, "Seahawk Joins Surface Development Squadron One," *Leidos*, May 27, 2021. www.leidos.com.

8. Quoted in "DARPA Christens Unmanned Ship Aimed at Revolutionizing Naval Capability," *DARPA*, August 11, 2025. www.darpa.mil.

FOR FURTHER RESEARCH

BOOKS

Cynthia Kennedy Henzel, *Space Defense Systems in the Military.* BrightPoint Press, 2027.

George Anthony Kulz, *Robotics in the Military.* BrightPoint Press, 2027.

Carl Meister, *Military Machine Fact Frenzy!* Capstone Press, 2026.

INTERNET SOURCES

"Marine Corps Launches New Drone Training Program," *US Department of War*, 2025. www.war.gov.

"Space Place: What Is a Laser?" *NASA*, 2025. https://spaceplace.nasa.gov.

Aušrine Vaiciute, "Can Sound Be Used as a Weapon?" *Focusonics*, 2025. www.focusonics.com.

WEBSITES

Defense Advanced Research Projects Agency (DARPA)

www.darpa.mil

The DARPA website provides information about cutting-edge technology being developed by the Defense Advanced Research Projects Agency.

Military.com

www.military.com/equipment/weapons

The Military.com website provides in-depth information about many military weapons.

Warfare History Network

https://warfarehistorynetwork.com

The Warfare History Network features details about the weapons and strategies used during the Vietnam War (1955–1975).

INDEX

IMAGE CREDITS

Cover: Samuel King Jr./DVIDS
5: Spc. Alejandro Carrasquel/US Army/DVIDS
7: © Parilov/Shutterstock Images
8: Sgt. Eli Baker/US Army/DVIDS
10: Brandy Plucker/US Navy/DVIDS
13: © emerald_media/Shutterstock Images
17: Kayla Cosby/US Army/DVIDS
18: Capt. Austin May/DVIDS
21: Lance Cpl. Joshua Simmonds/US Marine Corps/DVIDS
22: © TSViPhoto/Shutterstock Images
25: Mark Schauer/US Army/DVIDS
26: Spc. Kyle Goines/US Army/DVIDS
29: Darrell Ames/US Army/DVIDS
33: Lance Cpl. Andrew Huff/US Marine Corps/DVIDS
34: Sgt. Trent A. Henry/US Marine Corps/DVIDS
37: Giancarlo Casem/DVIDS
41: © Allora Empire Art/Shutterstock Images
42: Spc. Marques Martinez/US Army/DVIDS
44: Matthew Ryan/Bell/DVIDS
47: © Photomarine/Shutterstock Images
48: © Aleksandr Merkushev/Shutterstock Images
50: Petty Officer 3rd Class Kristopher Wilson/US Navy/DVIDS
51: © igorwall/Shutterstock Images
53: © Aerial-motion/Shutterstock Images
55: Drew Verbis/Northrop Grumman//DVIDS
56: Petty Officer 2nd Class Jesse Monford/US Navy/DVIDS
57: DVIDS

ABOUT THE AUTHOR

Cynthia Kennedy Henzel has degrees in education and geography. She has written more than a hundred books for young people. These include biographies, fiction, and nonfiction on subjects such as geography, science, culture, and history.